How to Become a Fossil

Anne Lamb

Consultants

Matthew T. Miller
Museum Specialist
National Museum of Natural History

Cheryl Lane, M.Ed.
Seventh Grade Science Teacher
Chino Valley Unified School District

Michelle Wertman, M.S.Ed.
Literacy Specialist
New York City Public Schools

Publishing Credits

Rachelle Cracchiolo, M.S.Ed., *Publisher*
Emily R. Smith, M.A.Ed., *SVP of Content Development*
Véronique Bos, *VP of Creative*
Dani Neiley, *Editor*
Robin Erickson, *Senior Art Director*
Jill Malcolm, *Graphic Designer*

Smithsonian Enterprises

Avery Naughton, *Licensing Coordinator*
Paige Towler, *Editorial Lead*
Jill Corcoran, *Senior Director, Licensed Publishing*
Brigid Ferraro, *Vice President of New Business and Licensing*
Carol LeBlanc, *President*

Image Credits: p. 4 John Sibbick/Science Photo Library; p. 17 (bottom) Alamy; p. 18 Universal Images Group via Getty Images; p. 20 (top) South China Morning Post via Getty Images, (bottom) to follow; all other images from iStock and/or Shutterstock

TCM | Teacher Created Materials

5482 Argosy Avenue
Huntington Beach, CA 92649
www.tcmpub.com
ISBN 979-8-7659-6893-2
© 2024 Teacher Created Materials, Inc.
Printed by: 51497
Printed in: China

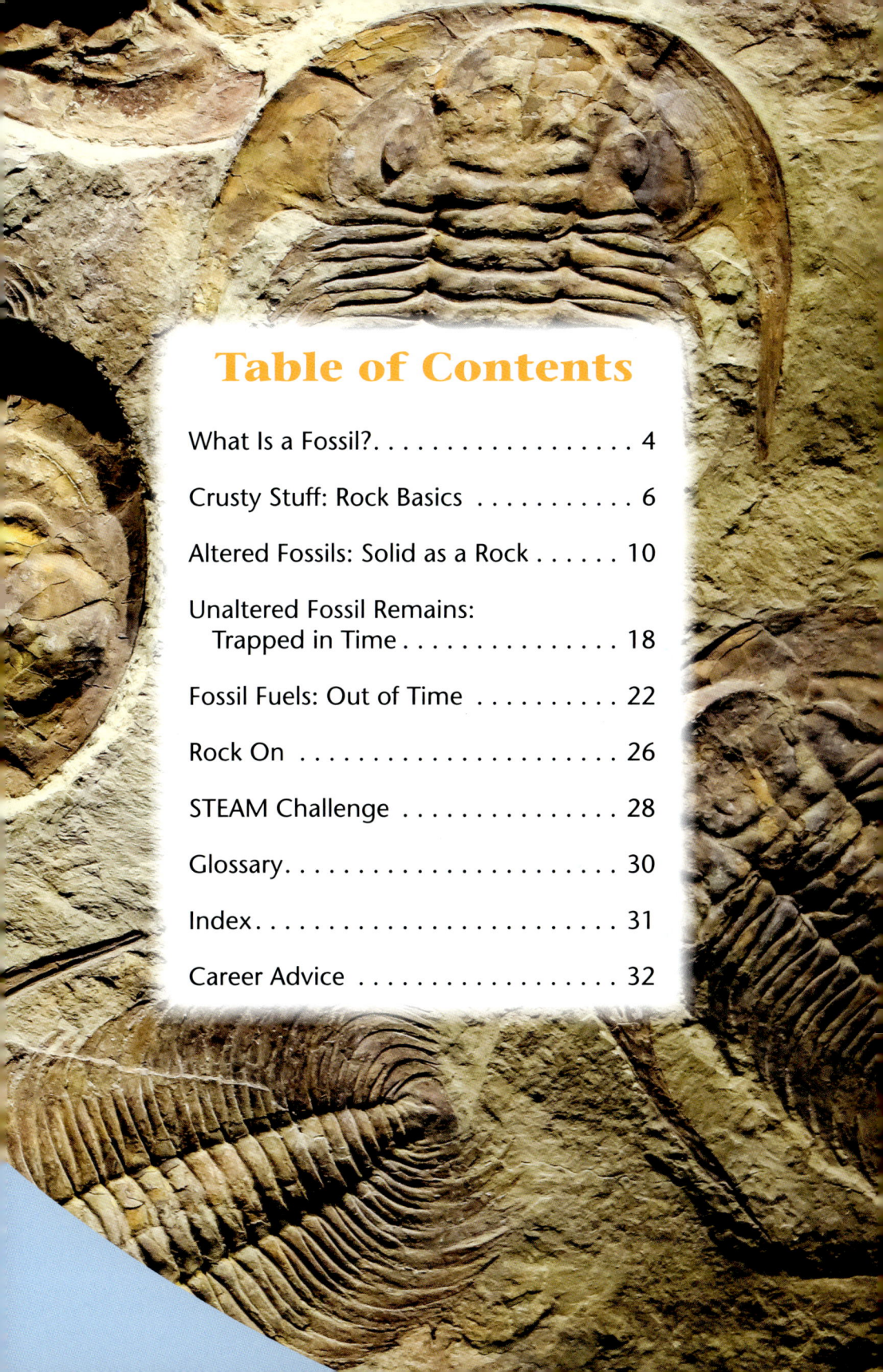

Table of Contents

What Is a Fossil?

A fossil is the preserved remains of a **prehistoric** creature or plant that is at least 10,000 years old. These remains have been partly or totally replaced by minerals. Preserved traces of prehistoric creatures, such as footprints, are also fossils. Fossils have been found all over the world. Some fossils are put on display in museums, giving us a glimpse into the past.

fossilized leaf

Scientists use the word **fossilization** to describe how fossils form. There are many fossilization processes. They require certain conditions and create different types of fossils. For example, one type of fossilization creates leaves printed on rocks. Another creates preserved animal teeth or bones.

fossilized megalodon teeth

Drawing from the Past

ARTS

Some artists use their skills to paint pictures of Earth's past. They reference fossils to create images of what prehistoric animals and their ecosystems may have looked like. This type of art is called *paleoart*. It can be shown in 2D or 3D forms.

Some fossils are a hidden part of everyday life. Have you ever taken a bus? What about cooked on a gas stove? Fossil fuels, such as coal, oil, and natural gas, make those activities possible. However, they are a limited resource. **Extracting** them has a huge environmental impact.

People study fossils to learn about the past. Fossils tell us what kinds of animals and plants lived on Earth long ago. Also, fossils offer clues about Earth's geological cycles and climate changes.

Want to see how a fossil is made? All it takes is **organic** remains, the right set of geological conditions, and at least 10,000 years. But no need to wait—Earth already has a variety of fossils to explore!

fossilized dinosaur skeleton

Crusty Stuff: Rock Basics

Before learning about how fossils are made, it helps to know about Earth's outermost layer—the crust. Some of the same processes that help make fossils also help shape Earth's crust.

Earth's crust is made of rocks, and rocks are made of minerals. But this doesn't mean the crust never changes. In fact, Earth's crust is constantly recycling its materials. This recycling process is known as the rock cycle. Earth's crust has three types of rocks: igneous, sedimentary, and metamorphic. They're all part of the rock cycle.

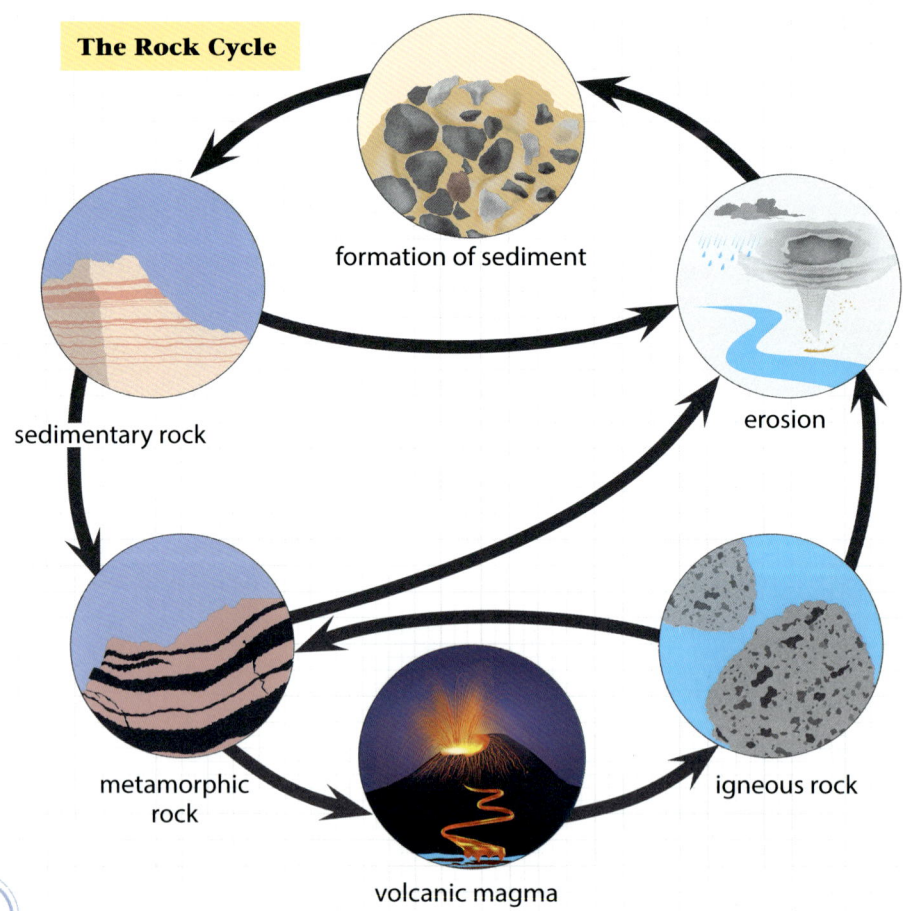

The Rock Cycle

formation of sediment

sedimentary rock

erosion

metamorphic rock

volcanic magma

igneous rock

Basalt is a type of igneous rock.

Igneous rocks form when magma, the liquid rock under Earth's surface, cools and hardens. Igneous rocks can form in Earth's crust or at volcanoes on the surface.

Sedimentary rocks are made of **sediment**. Sediment is Earth's greatest mash-up. It's made up of rocks, soil, minerals, organic matter, and other debris. Some sedimentary rocks form when sediments are pressed together over time. Other sedimentary rocks form when minerals that were dissolved in water become solid.

Marble is a type of metamorphic rock.

Metamorphic rocks form when existing rocks, such as sedimentary or igneous rocks, get buried deep in Earth's crust. There, the increased heat and pressure changes the chemicals in the old rocks. The old rocks become new metamorphic rocks.

Earth's crust is a mixture of old rocks, new rocks, and old rocks that become new rocks. This is great news for making fossils. Without the rock cycle, many types of fossils would not exist!

FUN FACT

Scientists can use technology to determine the age of rocks. The oldest rock that has ever been found on Earth is more than four billion years old. It was found in northwestern Canada.

Sedimentation

The rock cycle depends on many forces. Some of these forces are volcanic eruptions, tectonic shifts, erosion, and **sedimentation**. Sedimentation is one of the forces always at work in recycling materials on Earth. It's also an important part of creating many types of fossils.

The process first starts with erosion. Erosion occurs when wind, rain, or ice breaks down rocks and soil into smaller bits of sediment. Then, water and ice move the sediment. Rivers, oceans, lakes, glaciers, and streams all carry sediment from one place to another. When water drops that sediment, it's known as sedimentation.

Sometimes, this process forms layers of sediment. In some places, sediment layers have been preserved for millions of years. Scientists study the layers to understand what Earth was like when the layers were created. But sediment does so much more!

With lots of pressure and time, some sediments become rocks. Sandstone and limestone are two types of sedimentary rocks. Sediments can also become fossil fuels. Crude oil and natural gas are two types of fossil fuels. And sometimes, sediments are deposited around or inside the remains of a plant or an animal. This creates fossils.

sedimentary rocks

fossils in sandstone

These sedimentary rocks in Utah have visible layers.

Altered Fossils: Solid as a Rock

Altered fossils are made of rock or are found in rocks. As these types of fossils formed, the organic material of the animals or plants changed. Sediment, heat, pressure, and long amounts of time are needed to form altered fossils.

Body fossils and trace fossils are two types of altered fossils. Body fossils serve as records of the physical remains of plants or animals. Bones, shells, and teeth are examples of body fossils. Trace fossils record an activity, or a trace, of an organism's life. Footprints, droppings, nests, and burrows are all trace fossils. They provide evidence of living animals but do not include any body parts or remains.

SCIENCE

Prehistoric Hints

Scientists learn different things from body and trace fossils. Trace fossils help scientists learn how animals lived. Scientists can study how they moved, burrowed, or nested. Body fossils help scientists learn what animals looked like. Scientists can also study how the animals died.

dinosaur footprint

fossilized dinosaur eggs

carbonized fish fossil

Carbonized Fossils

Have you ever pressed a flower in the pages of a book? Creating carbonized fossils starts like this. But instead of paper pages, layers of sediment compressed prehistoric remains.

carbonized insect fossil

Carbon is an **element**. It exists in every living thing. In a carbonized fossil, high pressure and heat in Earth's crust flatten an organism's remains in sediment. After millions of years, only a carbon imprint of the organism is left behind in the stone. Some carbon fossils have preserved the soft tissue of organisms. These carbon fossils can show many details, such as fish scales or fern leaves.

Petrified Fossils

Petrified fossils are organic remains that have been turned into stone. **Permineralization** plays a large part in the creation of these fossils. This is a big word for a process that happens in tiny parts of organisms. During this process, minerals permeate, or spread, through organic matter. This process creates petrified fossils.

The process begins when a plant or animal is quickly buried under a lot of sediment. This might happen in a volcanic eruption or a flash flood. The quick burial slows decay. Then, mineral-rich water trickles down through the sediment. The water seeps into the remains. It enters the cells of the organism. As the water evaporates, it deposits minerals in the cells. This preserves the shape of the cells. Little by little, minerals replace most or all the organic material. This is why some petrified wood retains its growth rings. In these cases, some of the original matter still remains. It is preserved by the mineralized cells around it.

Petrified fossils show the internal shapes of plants or animals. The minerals that make up these fossils can tell scientists where the organisms died. For example, volcanic ash is full of **silica**. Fossils that contain silica may have formed after an eruption. And fossils that have **sulfur** typically come from watery environments. Different materials found in fossils tell the story of where a plant or animal once lived.

FUN FACT

Petrified Forest National Park is full of petrified wood. The park is located in northern Arizona. Visitors can walk on trails that weave through petrified stumps and logs. The park also has colorful badlands made up of rounded hills and cliffs.

The various colors in petrified wood are caused by different elements.

Mold and Cast Fossils

Mold and **cast** fossils know how to make a good impression! These fossils begin forming when the remains of an organism fall into soft sediment layers, such as those at the bottom of a river or seabed. The soft parts of the organism start to decay quickly. Any hard parts, such as a shell, are left behind. As more sediment covers the shell, it makes an impression in the sediment beneath it.

Over time, water continues to deposit sediment around the shell, encasing it. The hard shell takes a long time to decay. While the shell slowly decomposes, the sediment around it becomes harder and more compact. It begins turning into rock.

Eventually, the original material of the shell decomposes completely. It leaves behind an empty space inside that rock. The space is in the exact shape and size of the shell. This is a mold fossil. Mold fossils can form from bones and leaves, too.

mold fossil

Cast fossils are formed with the help of mold fossils. Sometimes, water rich in minerals or sediment seeps into a mold fossil. The water deposits the minerals into the mold. Little by little, the minerals harden and gradually fill the empty space. Eventually, the space is completely filled with a new rock in the same shape and size of the old shell. It's a copy of the old shell. This is known as a cast fossil.

cast fossil

mold fossil

ENGINEERING

Fossil Feats

Real dinosaur fossils are very old, fragile, and heavy. Putting the real fossils on display in museums can put them at risk of breaking. So, some of the dinosaur bones on display are casts of fossils. They are typically made of lightweight fiberglass or plastic. Using casts allows museums to fill in missing bones and safely display exact copies of fossils.

triceratops fossil at a museum

Finding Fossils

Removing an altered fossil from its surrounding rock is tricky work. In the field, paleontologists might use shovels and chisels to remove big chunks of rock from around a fossil. Once the fossil is in the lab, scientists use smaller, specialized tools. These tools allow them to carefully remove any remaining rock from a fossil. Many fossils are still prepared and studied this way. But new, cutting-edge tools are changing the ways scientists look at ancient remains of life.

A paleontologist works on a dinosaur fossil.

FUN FACT

Some scientists study fossils to learn about the future. One researcher named Selina Cole works toward this goal. She studies crinoids. These animals are related to sea stars and sea urchins. Cole examines mass extinction events of the species in the **fossil record**. This information is useful in learning which modern species may be at risk of extinction.

crinoid fossil

A researcher studies the CT scan of a *T. rex* skull.

Sometimes, scientists don't even need to remove the fossil from a rock to study it. Instead, they can use detailed CT scans of a rock. A CT scan requires a machine that uses computer technology to take a lot of X-ray-like images at once. Scientists convert the scans into 3D images of the bones. Then, they feed those images into a 3D printer and study the bones on a computer screen. They can also print copies of the bones to examine.

Other scientists use advanced computer programs to help them analyze fossils. They use a type of artificial intelligence, or AI, to speed up their work. Researchers have trained the AI program to recognize fossils in CT scans of rocks. The AI program analyzes the scans much more quickly than human researchers can.

In the future, technology will continue to improve. Scientists will likely get to use new tools in the field. There is so much more for them to learn about how to find and study every type of fossil.

Unaltered Fossil Remains: Trapped in Time

While many fossils are stone replicas of bones, some fossils preserve parts or whole bodies of organisms. These types of fossils are called unaltered fossil remains.

An ant trapped in amber is an example of an unaltered fossil remain. So is a mammoth frozen in **permafrost**. Bones recovered from a tar pit are also unaltered fossil remains. In all these cases, the chemical makeup of the animals has not changed since they died.

How are these examples different from altered fossils? In altered fossils, the original bones are long gone. Unaltered fossil remains still have the original bones. They have been preserved for tens of thousands, or even millions, of years!

In 1977, Russians found this baby mammoth frozen in permafrost.

frog in amber

insect in amber

Amber Fossils

Amber is the fossilized resin, or sap, from prehistoric trees. Resin is a thick and sticky material. It drips and flows from a tree's bark, covering and trapping all kinds of things. Resin can cover insects, plants, soil samples, and even small reptiles.

Eventually, the resin blob falls to the ground. Gradually, it is buried and pushed deeper into Earth's crust. As the resin is exposed to heat and pressure over millions of years, its molecules change. It hardens into amber. This process is known as **polymerization**. The amber perfectly preserves the organisms inside it so they do not decompose.

resin

Frozen Fossils

Unaltered fossil remains can also be formed through the process of freezing. This process shows another way that water helps make fossils. Frozen fossils can be found in certain places around the world where there are layers of permafrost in the ground. This permanently frozen soil can be a good environment for fossil formation. That's because decay is slowed when a dead animal's body freezes in ice or permafrost. This preserves the skin, fur, and internal organs for tens of thousands of years. In some frozen fossils, an animal's last meal has even been preserved in its stomach! A frozen fossil can show how a prehistoric animal looked and behaved when it was alive.

frozen baby mammoth

Fossil Forensics

In rare cases, scientists are able to extract DNA from the bones or teeth of well-preserved frozen animals. They can use DNA to learn more about the animals and how they lived. But if the bones or teeth are extremely old or not well-preserved, it is much harder to find and extract DNA. That's because DNA is a fragile molecule that degrades over time.

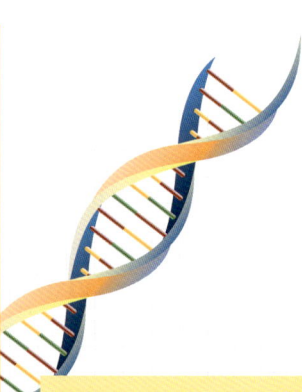

DNA is a tiny molecule made of two linked strands.

Tar Fossils

Tar pits form when crude oil in Earth's crust seeps up and pools on the surface. Over time, some of the oil evaporates. This creates fields of super sticky **asphalt**, which are known as asphalt seeps. They are also nicknamed "tar pits." These pits trapped and preserved prehistoric plants, insects, and animals. Often, when an organism became trapped in the asphalt, its soft tissues decomposed. Then, the asphalt preserved the hard parts, such as bones or shells.

lake pit at La Brea Tar Pits

At La Brea Tar Pits in California, scientists discovered fossils of large Ice Age mammals. These included mastodons, dire wolves, and saber-toothed cats. They also discovered beetles and birds. But one thing that will never be found at La Brea is dinosaur bones. The dinosaurs became extinct millions of years before the tar pits formed!

bison skeleton at La Brea Tar Pits

Fossil Fuels: Out of Time

Fossil fuels are formed from the fossilized remains of plants and animals. Crude oil, natural gas, and coal are all fossil fuels. Crude oil is used to make fuels and products, such as gasoline, plastics, and medicine. Natural gas is used for cooking, heating, and powering some vehicles. Coal powers factories when it's burned. Although fossil fuels are used regularly on Earth, they are nonrenewable resources. This means when Earth runs out, people won't be able to make more. Fossil fuels can take thousands or millions of years to form.

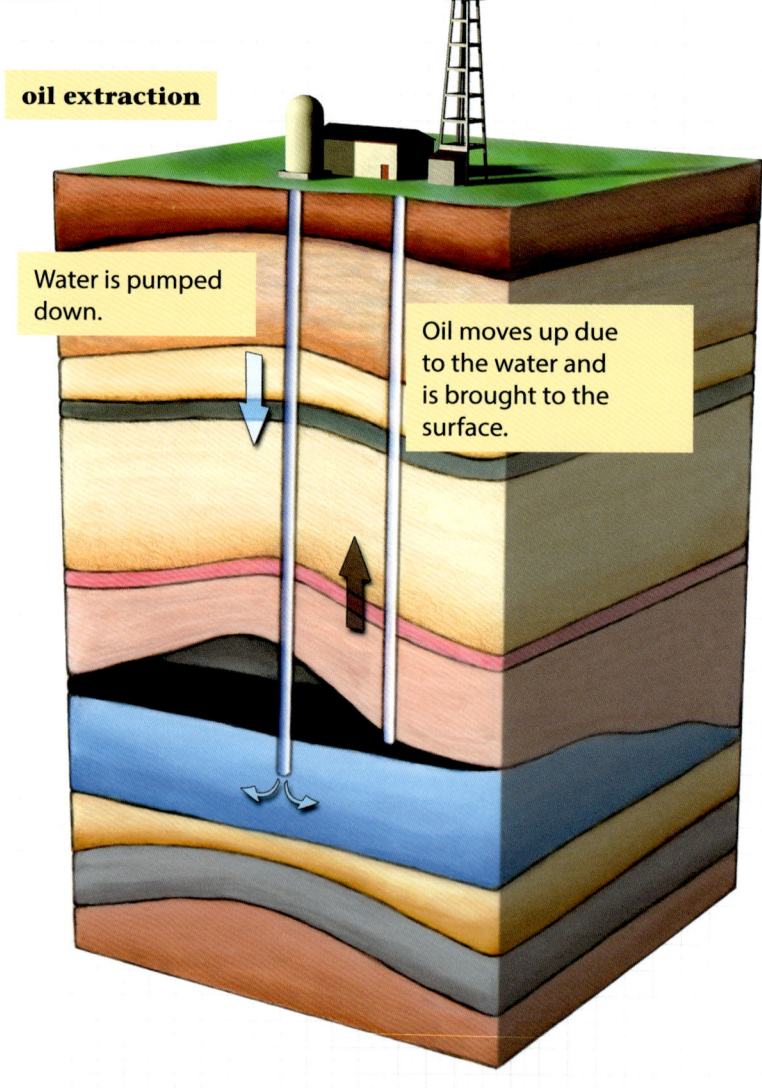

oil extraction

Water is pumped down.

Oil moves up due to the water and is brought to the surface.

rig for oil well drilling

Petroleum

Petroleum comes in a few forms that all have the same origin story. Their story begins hundreds of millions of years ago with the remains of ancient marine plants and animals. These remains formed layers of organic sediment on the sea floor. More sediment covered the organic layers. This pushed the organic material deep into the ground where it was compressed and heated.

Over millions of years, the extreme pressure and heat changed the organic layers into petroleum. The deeper, hotter layers of petroleum became natural gas. The cooler layers closer to the surface became crude oil. Bitumen, a semi-solid form of petroleum, was formed as well.

Petroleum is used around the world. But every step in mining and using petroleum harms the environment. Drilling wells to access gas and oil risks spills and leaks. Refining petroleum creates harmful chemicals. Burning petroleum-based fuels creates **greenhouse gases** that cause climate change.

Pumpjacks bring oil to the surface.

Coal

Coal's story begins in peat bogs. Peat bogs are wetlands made from compressed layers of dead, decaying plants. These dead plants include prehistoric mosses. Peat bogs are also **carbon sinks**. They contain lots of carbon dioxide and can take thousands of years to form. But when some peat bogs were buried under sediment, they compressed even more. The pressure and heat increased. Over millions of years, the carbon in the peat changed into hard coal.

peat bog

Coal has been used as a resource by humans since cave dwellers first burned it for heat. In the 1800s, coal fueled boats and trains. Today, coal fuels power plants and factories. But burning coal creates harmful chemicals that contribute to acid rain, smog, and climate change.

MATHEMATICS

Running on Empty

People have known that fossil fuels are nonrenewable for years. However, the use of fossil fuels continues to rise. At our current rate of use, some studies say that Earth will run out of all fossil fuels by 2090!

power plant

Renewable Alternatives

Fossil fuels are nonrenewable resources. Human use of them causes damage to the environment. Often, this damage is permanent, or it can last for many years. To protect life on Earth, scientists think that people will need to stop using fossil fuels. Instead, they'll need to use more renewable energy sources. These include wind and solar power. Experts also say that people should work on reducing the amounts of energy they use.

Wind turbine blades capture wind energy, which gets converted into electricity.

Solar panels absorb sunlight and convert it into electricity.

Rock On

Becoming a fossil is hard work. It takes a lot of time and materials to preserve the remains or traces of an organism. Also, very specific conditions need to occur for fossils and fossil fuels to form. For example, pressure and temperature have to be just right.

Some fossils are altered fossils. These include fossils made out of rock or found in rocks, such as carbonized fossils, petrified fossils, and mold and cast fossils. Other fossils are unaltered fossil remains. These include organisms preserved in ice, fossilized amber, or tar pits.

While fossils can form in many ways, they are surprisingly rare. Fossils only record a fraction of all life that has ever lived on Earth. And only a tiny percentage of all fossils are discovered!

fish fossil

insect in amber

Preparing fossils involves careful work and special tools.

But some fossils are part of daily life. Crude oil, natural gas, and coal are fossil fuels. They come from the fossilized remains of prehistoric organisms. People use fossil fuels to generate heat and energy. They also use them to make things, such as plastics and medicines. But someday, Earth may run out of fossil fuels if changes are not made to slow usage.

Scientists are using new technology to find and study fossils. Advances in DNA processing, computer imaging, and AI give scientists new ways to study fossils. Fossils are relics of Earth's past that hold important information about Earth's future. With every new discovery, fossils preserve their place in history.

STEAM CHALLENGE

Define the Problem

Your local natural history museum has run out of educational items to include in their museum store. They have asked you to create a new educational game that is enjoyable, teaches players about fossils, and features models or puzzles. They have suggested making a board game, but they are open to other game ideas if you have them.

Constraints: You may only use the materials provided to you.

Criteria: Your game must have a fossil theme. Players must answer questions about fossils at some point in your game. Players must also be required to build models or complete puzzles about fossils to win the game. Your game should have clear directions that can be followed by other students your age.

Research and Brainstorm

What types of fossils are there, and how are they formed? If you are creating a game board, how will players move from start to finish? How can players learn about fossils as they play your game? When in games do players have to answer questions? Which games do you enjoy playing, and what ideas from them can you use in your own fossil game?

Design and Build

Sketch two or more designs for your game. Label the parts and the materials. Write sample directions. Choose the design you think will work best. Then, build your game.

Test and Improve

Share your game with others. Explain the directions, and have them play the game. Then, ask them for feedback. Does your game have a clear fossil theme? Do players have to answer questions about fossils at some point? What about your design is working well? How can you improve it? Modify your design and rebuild it as needed. Reassess how well it meets the criteria.

Reflect and Share

What about this challenge was the most difficult? Do you think this game would sell well at a gift shop in a natural history museum? What else could you add to your game to make it more interesting or educational? Why do you think people choose to study and search for fossils?

Glossary

asphalt—a thick, sticky brown/black mineral substance

carbon sinks—natural environments that absorb carbon dioxide from the atmosphere

cast—something that is formed by a liquid substance hardening over time in a mold

DNA—the information inside the cells of living things that acts like instructions for how to make that organism's body; abbreviation for deoxyribonucleic acid

element—any of more than 100 fundamental substances that consist of atoms of only one kind

extracting—removing something from something else

fossilization—the process of a plant or an animal becoming preserved

fossil record—record of all documented fossils that shows the history of life

greenhouse gases—gases that trap heat in Earth's atmosphere and contribute to the greenhouse effect

organic—of, relating to, or obtained from living things

permafrost—a permanently frozen layer below the surface of the earth in cold regions

permineralization—a type of fossilization in which minerals in water fill all the spaces in the remains of a plant or animal

polymerization—a chemical reaction in which larger molecules form from the combination of two or more small molecules

prehistoric—of, relating to, or existing in times before written history

sediment—solid material (such as stones and sand) deposited by water, wind, or glaciers

sedimentation—the action or process of forming or depositing sediment

silica—a compound that consists of the dioxide of silicon and occurs in various forms (as in quartz, opal, and sand)

sulfur—a nonmetallic chemical element that is an essential element for all life

Index

CAREER ADVICE
from Smithsonian

Do you dream about finding or working with fossils?

Here are some tips to keep in mind for the future.

"You can learn a lot by going outside and observing nature. All modern life has fossil ancestors, so by learning about the present you also learn about the past. Take notes, photographs, and drawings as you go. Slow down and look for small details. You will soon find yourself recognizing features in living animals and plants that you also see in fossils!"

– Michelle Pinsdorf, Research and Collections Assistant, Vertebrate Paleontology, National Museum of Natural History

"Be brave and ask questions. Explore your local museums and learn about what kinds of fossils are found where you live."

– Myria Perez, FossiLab Manager and Educator, National Museum of Natural History